The
Collapse
of 2020

Kirkpatrick Sale

TABLE OF CONTENTS

I.

THE INCIPIENT BET

IN THE FALL of 1992, after I finished my second round of speeches for the Quincentennial of America's "discovery," drawing on my book about Columbus that had been published in 1990, I started on a book on a subject that had long fascinated me and I knew was just as timely as the Columbus book had been. It was a history of the Luddites, that group of workers in the mid-country of England who had resisted the onslaught of the Industrial Revolution in the early 1800's by breaking offensive machines and had thus become the prototype of people who supposedly always resisted technology.

There had been no book published in this country telling the full story of the Luddites and only two books reprinted from meager English sources, so I knew that a book just now, when the internet was just beginning to be a means of communication beyond the scientific community that had developed it in the 1980's, and computers were becoming ubiquitous, would certainly be needed.

And so I went to work, impelled by the need to give a sympathetic telling of the long-forgotten cause and a desire to put the Luddite tale of resistance to the machine before a public, given the increasing impact of the computer-created second Industrial Revolution then upon daily life, that might well might be in the mood to wonder about people who had suffered from the first. The book came out in the spring of 1995 to some modest praise and, from the mainstream press, some perplexity that anyone would want to bring up those people again, particularly because they had failed in their attempts to stem the tide of progress. And among a small but growing number of people, critics of the modern version of that tide who were beginning to call themselves "neo-Luddites," it became what one of them called "the bible for our movement."

I was not of course neutral in my retelling of the Luddite tale nor in using it as a theoretical guide in analyzing and castigating our contemporary economy. My analysis of this second Industrial Revolution ended this way:

> Neither the means nor ends of this revolution make any pretense to nobility, nor does it claim to have a vision much grander than that of prosperity, perhaps longevity, both of which have seriously damaging environmental consequences. By and large its adherents would disclaim grand-sounding motives or methods, arguing them irrelevant: enough that progress may be presumed meritorious and material affluence desirable and longer life positive. If industrialism is the working-out in economic terms of certain immutable laws of science and technology, of human evolution, the questions of benign-malignant hardly apply. Those who want to wring hands at this inequity or that injustice are essentially irrelevant, whether or not they are correct: the terms of the game are material betterment for as many as possible, as much as possible as fast as possible, and matters of morality or appropriateness are really considerations of a different game.

Viewing the issue on its own terms, the question would then seem to come down to the practical one of the price by which the second Industrial Revolution has been won and whether it is too high, or soon will prove to be. The burden of this chapter is that there are no two ways of answering that question: the price, particularly in social decay and environmental destruction, has been unbearably high, and neither the societies nor ecosystems of the world will be able to bear it for more than a few decades longer, if they have not already been overstressed and impoverished beyond redemption.

To put it simply, modern technology enables us to do the bad things we're doing faster and more efficiently than ever before.

You can imagine what the reaction of the champions of that Revolution must have been. Various Silicon Valley pundits were quick to denounce such heresy and among the crowd in California who had always looked favorably on technology—the patrons of the Whole Earth Catalog (an "access to tools" that prefigured the web) in the 1960's—there was a feeling that a counterattack ought to be mustered as soon as possible.

It was in just such a role that one day in 1995 Kevin Kelly of *Wired* magazine in San Francisco called me in New York and asked for an interview. I had never heard of either one, but I agreed, went out and bought a couple of copies, and discovered that *Wired* was wide-eyed and indiscriminate in its love of technology and apparently blind to whatever downside it may have; Kelly, 42, was the executive editor. A few days later he came to my office in my apartment in New York's Greenwich Village and set up a recorder to interview me.

He made no attempt to pretend impartiality, his questions were hostile from the first. But he let me lay out my ideas about the Luddites, about the relevance of their resistance to the machine, about the dangers that machines far more powerful than theirs were doing today and were likely to do in the coming decades. Kelly some years later said, "I was irked by his assertions of the coming collapse of civilization" and he challenged me to spell out in exactly which ways modern technology would likely cause the collapse of civilization as we know it—and when that would happen.

I was stumped for a bit because first of all it was hard to isolate just a few ways in which such a monumental catastrophe would come about and secondly I had given absolutely no thought as to when. Finally I blurted out "2020" because it had a nice familiar ophthalmological ring and at that time it seemed far, *far* away, and I tried to lay out in as succinct a way as I could how it could happen.

First, an economic collapse. I posited that it might take the form of a worldwide currency devaluation, in which the dollar loses its standing as the world's reserve currency and becomes effectively worthless even in this country, and a global stock-market crash and depression.

Second, a political collapse, with upheavals both within nation-states and between. I saw the collapsed economy leading to maybe the bottom fifth of society in the developed world, no longer bought off with alcohol and drugs and celebrity and consumerism, rising up in rebellion and creating havoc and disarray throughout; at the same time a similar rebellion of the poor nations, no longer content to take the crumbs from the table of the rich, and simultaneously fighting violent guerrilla wars and flooding into the developed nations to escape their misery.

And finally, perhaps over-arching, an environmental collapse, in which global warming and ozone depletion, for example, made some areas like Australia and Africa unlivable and caused ice packs to melt, and old diseases, released from melting ice and deforested swamplands, mixed with new and spread deadly infections to all continents.

In truth, I had read a good deal on prognostications about future calamities—the Union of Concerned Scientists, for example, had issued a "warning to humanity" about "vast human misery" in 1992—but I had not before ever formulated them, never tried to be specific about the ways they would come about. So this was a first stab, quickly contrived and very imperfect, at what kinds of things I thought might befall the earth—nothing like what I could formulate just a decade later in another book.

It was enough for Kelly. With some sense of triumph, he whipped out a check for a thousand dollars and said, "I bet you US$1000 that in the year 2020, we're not even close to the kind of disaster you describe." He had obviously planned to maneuver me into this kind of challenge from the beginning. "We won't even be close. I'll bet on my optimism."

Well, a thousand dollars in those days was a lot of money, period, and especially for me, a serious independent writer whose book sales were respectable but not spectacular. A thousand was about all I had in my checking account. But then I thought: what the hell, if I'm right, nobody is going to win a bet in 2020, and if I'm wrong the dollar still won't be worth very much by that time.

I reached into my desk for the checkbook and quickly wrote out a $1000 check, dated March 6, 2020. I handed it to Kelly, shook his hand, and he said, "Oh, boy, this is easy money—but, you know, besides the money, I really hope I am right."

After just a moment I said, "I hope you are right, too."

We are now approaching that date—March 2020—and I thought it would make sense to see how close Western civilization has come to the predicted collapse in the year I predicted.

Obviously because you are reading this, and from a book published by electronic and mechanical machinery in workable order, the collapse has not occurred, or at least not as severely as I had thought. Therefore, however dire our conditions are now—and I would say that they are dire indeed—they do not add up to a complete collapse…or not yet.

So let me just admit that I was wrong. But—as you will see in the course of this book—not by much. And not totally. So that means we have time *right now* to prepare for the coming catastrophe and create strategies that might permit the survival of some in some places of refuge.

And though I treat the elements of collapse separately in the pages that follow, it is necessary to understand that they are all intimately interlinked, so that the failure in one area causes a failure in another—as for example, when increasingly hot weather causes crop failures causes famines causes immigration causes border invasions causes warfare. As we are interlinked in a global ecosystem, so we are interlinked in tragedy.

Jem Bendell worked for 25 years on something he called "sustainable development"—the idea of doing business without depletion or degradation of natural resources and ecosystems—and he was successful in finding clients like international businesses, investment funds, U.N. agencies, and the like. He wrote five books, including one in 2011 whose title suggests the sort of philosophy he was invoking: Evolving Partnerships: A Guide to Working with Businesses for Greater Social Change. *When the University of Umbria was formed in 2007, he was invited there as "Professor of Sustainability and Leadership," a position he holds today.*

But in 2018, after studying the ramifications of climate change he had something of an epiphany, deciding that his lifework had been essentially futile: "The whole field of sustainable development research, policy and education…is based on the view that we can halt climate change and avert catastrophe," he wrote in July. "By returning to the science, I discovered that view is no longer tenable," and there was very little prospect that any kind of social change would save the world from an imminent environmental disaster and "the likelihood of near term societal collapse." By that, he said, "I mean that within ten years, in whatever society we are living in, that we will find ourselves in a situation where our normal means of income, sustenance, security, pleasure, identity and purpose all disappear." Indeed, "the evidence before us suggests that we are set for disruptive and uncontrollable levels of climate change, bringing starvation, destruction, migration, disease and war."

Since Bendell sent out this alarm, dozens of other scientists have come forth to agree with his essentials, and most particularly the part about humankind's inability to make enough serious changes to the earth's environment in a short enough time to avert catastrophe. The levels of pollution reduction set by the Paris accord of 2015—to start by this year (2020) to reduce greenhouse gasses enough to meet a temperature limit of 3.6 degrees Fahrenheit—is simply laughable nonsense to anyone who has seen how high the earth's temperature has been in the last decade and how steadily the emission of CO2 in the atmosphere continues to increase every year, 2019 being the worst. Bendell puts it this way:

"*Important steps on climate mitigation and adaptation have been taken over the past decade. However, these steps could now be regarded as equivalent to walking up a landslide. If the landslide had not already begun, then quicker and bigger steps would get us to the top of where we want to be. Sadly, the latest climate data, emissions data and data on the spread of carbon-intensive lifestyles show that the landslide has already begun.*"

II.

ENVIRONMENTAL COLLAPSE: SELF-EXTINCTION

I HAVE ALREADY mentioned the Union of Concerned Scientists' warning in 1992 about "vast human misery" ahead, a caution to which the fast-engulfing computer revolution paid no heed.

It was followed in 2005 by a landmark Millennium Ecosystem Assessment, the most comprehensive environmental review ever done, that declared that two-thirds of the natural world that supports life on earth is being degraded by human pressure and "human activity is putting such a strain on the natural functions of Earth that the ability of the planet's ecosystems to sustain future generations can no longer be taken for granted." It too, clearly, went unheard.

In his June 2015 encyclical Pope Francis said bluntly: "Doomsday predictions can no longer be met with irony or disdain." So he thought. Or hoped.

In the fall of 2017, 25 years after the Union of Concerned Scientists' warning, a second group, this time of some 15,372 scientists from 184 countries, issued a new warning to humanity that if we do not make significant changes to our way of life immediately, we will no longer be able to ward off the inevitable precipitous decline of our planetary ecosystem, whose symptoms are getting steadily worse, as a result of our poor environmental stewardship. Still no significant response.

Although there are some who will deny the signs of ecological catastrophe and many who ignore these warnings as inconclusive, let me start out by listing some completely indisputable facts:

- The ten hottest years on earth, since quasi-global records began to be kept in 1850, have been between 2005 and 2019, and the hottest by far have been the last five years. The measurements are global and were monitored by National Aeronautics and Space Agency and National Ocean and Atmospheric Administration, plus multiple global stations in Europe, Asia, and the poles.

- While higher temperatures can be deadly to humans, as the heat waves in Europe and Asia demonstrated in 2016, 2018, and 2019 (the hottest year ever recorded), the most significant lasting effect is melting ice, at the poles, in Greenland, and at glaciers worldwide. In 2018, temperatures in the Arctic were 68F degrees above average and it was warming twice as fast as the rest of the world, with the result that sea ice was melting at an increasing rate and the area has been cut in half and the thickness has declined by 65 per cent in the last 30 years. In the Antarctic, where ice melt goes directly to increased ocean levels, the melting has gone three times as fast in 2018 and 2019 as in 2007 and sea level has increased by three times since 2012. Peter Wadhams, a leading Arctic scientist at the Polar Ocean Physics Group at Cambridge University and the author of *A Farewell to Ice,* has said bluntly that we can see "the Arctic death spiral."

- The first consequence of ice melting is a decrease in the albedo effect, the amount of sunshine that is reflected back into the atmosphere, and therefore an increase in the amount that comes in to warm the surface—by some calculations enough to be the equivalent of 25 per cent of the CO_2 in the atmosphere. And CO_2 is generally held to be a chief global warming gas, along with methane and, a new school of scientists is starting to believe, particulate pollution. The result: increasing temperature increases albedo increases temperature....

- A second consequence is a rise in sea levels. Arctic ice is already in the ocean, but Greenland and Antarctica are rocks with ice caps, so their ice-melts effect sea levels directly—and significantly; add in glacier melt and a rise from thermal expansion as the water gets warmer and the total indicates a striking rise of almost 3 inches from 2000 to 2018—and "the pace," warns NOAA, "is accelerating." Predictions vary, but a model assuming no significant reduction of CO_2 pollution figured a rise of 4 feet 9 inches during the 21st century, or over half an inch a year, and thus a rise of nearly 7 inches by 2030. Since at least 270 million people live in coastal areas, and two-thirds of the world's largest cities are coastal, such a rise would have serious flooding effects and many islands would be in peril. "The oceans...are in big trouble," said Michael Oppenheimer, lead author of the UN's Panel on Climate Change report to the UN summit in 2019, "and that means we're all in big trouble, too."

- A third result is extreme weather. Scientists used to be hesitant to blame global warming for weather conditions because the link was hard to prove, but all that changed in 2018 when a number of climatologists and meteorologists came out firmly for such an association. The increased heat, particularly in the Arctic, was shown to be responsible for vast jet stream alterations and resulting extreme weather events such as heat waves, increased rainfall and flooding, wildfire intensification, winter vortexes and low temperatures, hurricanes, and tornados.

- And perhaps most alarming of all, there is evidence of increased melting of the permafrost layers in the Arctic Ocean and the East Siberian Arctic Shelf, which is important because beneath them are massive amounts of methane gas, a far more potent greenhouse gas than CO2 and easily enough to tip the scales toward deadly warming levels. Methane levels have increased every year of this century and in 2018 reached record levels well over anything in the last 400,000 years. A team from the Arctic Methane Emergency Group, formed on Facebook in 2011, has reported that sudden eruptions of methane, which have happened 11 times in the geological past, including at the Paleocene-Ecocene extinction some 55 million years ago, can increase global temperatures by 5 degrees in the space of no more than 13 years. Not all Arctic scientists are as pessimistic as that, but veteran scholar Natalia Shakhova of the University of Alaska Arctic Research Center says such a pulse is "highly possible at any time."

That's just global warming. Let me add two other phenomena that serve to illustrate the catastrophic nature of the conditions we have created on earth.

The first is the assault on the oceans, not only from warming, which is considerable, but also from acidification, which increases the more carbon the oceans absorb, which some believe has increased by 50 per cent since the carbon era began. According to the World Wildlife Fund for Nature, half of marine vertebrates—whales, dolphins, fish turtles, seals—died out between 1970 and 2010 (other scientists say it is closer to three-quarters), and populations of tuna, mackerel, and bonito declined by 74 per cent; all indications are there's been an increase since then. For example, James Bradley, an Australian oceanographer, reported in August of 2018, in an article called "the end of the oceans," that a third of all large fish species in Australian waters disappeared between 2008 and 2018, and he feels the process is speeding up. With such species loss there is no way to make this up from elsewhere, because 90 per cent of the world's fisheries are already fished to capacity—or are overfished.

Another effect of acidification is the eradication of oxygen, which the UN 2019 climate report said had declined by up to 3 per cent over the last 50 years. The starkest effect of this is an increased number of "dead zones" in the ocean, and Lee Kump, Professor of Geoscience at Penn State University and long a student of ocean systems, has indicated that in such zones, "where the drop of oxygen is the combined effect of the warming, the small fish die out, unable to breathe, which means oxygen-eating bacteria thrive, and the feedback loop doubles back." And it means that plankton and other basic feed stocks for larger fish die out steadily. Plankton has declined by at least 40 per cent since 1950 and some dependent fish such as cod in the Atlantic and blue fin tuna in the Pacific among others have both seen declines.

Another basic ocean element for a great number of species are coral reef systems, and these have been hard hit by warm ocean temperatures and acidification in tandem. John "Charlie" Vernon—his nickname is from Darwin—is a legendary reef expert with 45 years at the Australian Institute of Marine Sciences and in late 2018 took some divers on a trip to the Great Barrier Reef off the Australian northeast coast. Referring to evidence that 30 per cent of the reef had perished in 2016 and another 20 per cent the following year, he said such "absolute massive death," not only the Great Barrier but reefs almost everywhere else, marks "the beginning of a planetary catastrophe."

The second phenomenon—and this is the overarching, the inescapable one—is the extinction of species.

Other than humans, that is. Obviously we are dependent—"utterly dependent" is the way Paul Ehrlich of Stanford puts it—on all the other species—animals, plants, insects, microorganisms—not just for food but for crop pollination and protection and the ongoing operations of multi-connected ecosystems. And they are going extinct at a rate not ever seen before by humans. A comprehensive study of 15,000 scientific papers issued by a United Nations task force in May 2019, the largest and most extensive ever undertaken, found that at least 1 million species of animals and plants are at severe risk of extinction. According to its chair, Sir Robert Watson, "The health of ecosystems on which we and all other species depend is deteriorating more rapidly than ever" and "we are eroding the very foundations of our economies, livelihoods, food security, health and quality of life worldwide."

Scientists reckon there have been five previous extinctions in the history of the earth, one from a meteor and resulting climate change, the others from climate change caused by greenhouse gasses. All evidence is that we are now in the sixth.

A paper published in July 2017 in the Proceedings of the National Academy of Sciences, by scholars from Stanford University and the University of Mexico, took an unusually blunt tone for a scientific treatise because they felt the situation was so drastic. After studying half the earth's vertebrate species, they concluded that "Earth was experiencing a huge episode of population declines and extirpations which will have negative cascading consequences on ecosystem functioning and service vital to sustaining civilization." And this sixth extinction, more severe than anyone had thought before, means a "biological annihilation" amounting to an "assault on the foundations of human civilization" and "a dismal picture of the future of life, including human life."

Quite a statement coming from scientists, most of whom, like the authors of the authoritative United Nations Intergovernmental Panel on Climate Change regular reports, seldom dare to go beyond "dangerous" or sometimes "alarming."

A new report the following year attempted to put quantitative figures to what the human imprint has meant for the world's biomass. Three biologists, two from the Weizmann Institute in Israel and one from the California Institute for Technology, determined that humans, .01 per cent of all life on earth, have annihilated 83 per cent of all wild animals, 80 per cent of marine mammals, 50 per cent of all plants, and 15 per cent of all fish. (A subsequent report, in October 2018, calculated also that between 45 and 76 per cent of insects have gone extinct.) Of what's left, livestock—domestic animals—accounts for 60 per cent and humans for 36 per cent, which leaves only 4 per cent for all wildlife. In other words, we have been so successful at birthing and developing our single bipedal species that we have virtually exterminated all the rest—so far, and we are proceeding to complete that at a pace so fast that it is beyond control.

Gobal warming and its many direful ramifications, dying oceans and their grim auguries, an extermination of life on a catastrophic scale: such then, is what the future surely portends. Whether that triggers the collapse in 2020, it surely suggests its imminence. Figure this: although there are perhaps ten trillions of planets in our galaxy, according to people who contemplate these things, and it just stands to reason that some of them surely have conditions similar to ours, and yet in the 2 million years that we have been humans we have no evidence that any of them has contacted us. This simple fact leads to one conclusion, one speculation. Perhaps it is that when any population develops something we would recognize as civilization, and that civilization develops industrialization to the point where it has or could create some form of space travel sufficient to make itself known to any other civilization, it collapses. Industrial civilization, in other words, is an inherently self-destructive system with limits beyond which it cannot survive, and utterly consumes itself like the self-burning tree of Gambia discovered by Mungo Park.

In 2013, after more than 30 years of environmental activism in the United Kingdom, Mike Ferrigan started a group on Facebook called the Near Term Human Extinction Support Group. (In the same year a Near Term Human Extinction Evidence *Group began.) The idea was to create a forum for people to exchange scientific papers and journalistic summaries of the environmental, political, social, and economic dangers that were imperiling the planet, and to share, in the site's words, "social, psychological, philosophic, spiritual, and emotional consequences of living with the knowledge that earth may soon be free of humans." At the end of 2019 almost 7,000 people had joined the group and it grows by about ten people a day. At the same time Ferrigan began a radio program called Extinction Radio that broadcasts regular interviews with scientists and activists in the field who confront the evidence for near-term extinction and work to dispel the complacency of those, within the scientific community and without, who believe that the actions taken now, or even those that might be taken if the world were to take the coming doom seriously, will avert the catastrophe.*

Near. Term. Human. Extinction.

In the prehistory of "civilization" many societies rose and fell, but few left as clear and extensive account of what happened to them and why as the twenty-first century nation-states that referred to themselves as Western civilization. *Even today, two millennia after the collapse of the Roman and Mayan empires and one millennium after the end of the Byzantine and Incan empires, historians, archeologists, and... paleoanalysts have been unable to agree on the primary causes of those societies' loss of population, power, stability, and identity.*

The case of Western civilization is different because the consequences of its actions were not only predictable, but predicted....While analysts differ on the exact circumstances, virtually all agree that the people of Western civilization knew what was happening to them but were unable to stop it.

--Naomi Oreskes and Erik M. Conway,
The Collapse of Western Civilization
(Columbia University Press, 2014)

III.

POLITICAL COLLAPSE: THE
CENTER CANNOT HOLD

THIS IS THE condition of the world in January 2020:

No less than 65 countries are now fighting wars—there are only 193 countries recognized by the United Nations, so that's a third of the world. These are wars with modern weapons, organized troops, and serious casualties—five of them, like Afghanistan, Libya, Syria, Somalia, and Yemen, with 10,00 or more deaths a year, another 15 with more than 1,000 a year—all of them causing disruptions and disintegrations of all normal political and economic systems, leaving no attacked nation in a condition to protect and provide for its citizens. From 2015 to 2019 more state-based conflicts were engaged in than at any time since World War II, with an estimated 1 million deaths in all.

In addition, there are at least 638 other conflicts between various insurgent and separatist militias, armed drug bands, and terrorist organizations, increasing each year as states fail or collapse completely. No continent is spared but Antarctica, and even there tensions are growing among nations eyeing its extensive underground mineral wealth, more precious as resources diminish elsewhere.

What has made the wars and internal disputes even more egregious as the years go on is that global warming has a direct effect on how societies function. Agriculture, of course, is impacted by higher temperatures, lack of rain, droughts, and wildfires, and crops have failed in many places over the last five years, including North and Central Africa, the Middle East, India, Pakistan, northern China, northern Europe, Argentina, Brazil, Central America, and even parts of North America. The collapse of Syria, for example, and subsequent civil wars were made more devastating if not directly caused by the drought of 2006-2011, in which 75 per cent of the farms failed and 85 per cent of the livestock died. And an official United Nations report in 2019, by 100 experts from 52 countries, warned that things will only get worse, with the world's land and water resources exploited at "unprecedented rates," threatening "the ability of humanity to feed itself."

One obvious consequence, beyond death, famine, disease and starvation, is, as the U.N. report's lead author says, "a massive pressure for migration," a desperate attempt to find some refuge and relief when homes have been destroyed and families are uprooted. According to the United Nations, in what I regard as a certain undercount, in 2019 there were 272 million migrants worldwide, up from 258 million in 2017, with the weather in 2019 causing more refugees even than warfare. (The unprecedented crisis at the U.S. southern border in 2019 is only one manifestation of the porous and chaotic collapse of boundaries across the Americas, Africa, and most of Asia.) And meanwhile, the International Committee of the Red Cross in 2018 estimated that more than 100,000 people are simply "missing," a figure it admits "represents only a fraction" of those who are unaccounted for by any government or organization.

Given the turmoil over wars and immigration threats, it is not surprising that half the world is without coherent government.

Organizations that track these things say that of the 174 covered nations, 76 are in various stages of collapse—that would be *43 per cent*—and that excludes a dozen smaller nations that are locked into autocracy and poverty. These include seven completely failed states—Congo-Brazzaville, Central African Republic, Syria, Yemen, Somalia, South Sudan, and Venezuela—and another seven that are on the edge—Guinea, Haiti, Iraq, Zimbabwe, Afghanistan, Chad, and the Sudan—plus 19 that are in an "alert" category, meaning that some but not all government functions have failed, 15 in Africa and 4 in Asia.

In other words, many political systems in the world have effectively collapsed, people are dispossessed and without governments, and almost everywhere else, including the U.S. and Europe, governments are severely strained and political rifts abound. The vote for Brexit in the U.K., the election of Donald Trump (and the subsequent attempt to overturn it), the turmoil that erupted in December 2018 in France and Belgium, the continued protests in Poland were all examples of the population of developed nations coming to see that the attempt to establish capitalist-led democracies in an internationalist arrangement of benefit to corporate and banking interests just was not working, and a rising segment of what were called "deplorables" in America did not want any longer to be powerless, manipulated, and disdained. These turmoils also demonstrated that the established powers in these countries, especially the U.S. and Britain, resisted all of these attempts to change the status quo and in effect ignored or thwarted the popular will—the developed world's form of the failed state. Those fissures have widened as the years have worn on, and as one astute observer, James Kunstler, put it in 2019, "The West is enduring paroxysms of political uproar and disenchantment."

Elsewhere in the world, violent anti-government protests continued to threaten established order throughout 2019. The most prominent, and deadly, were in Hong Kong, Chile, Lebanon, and Iraq, but there were others in India, Ecuador, Kazakhstan, Sudan, Saudi Arabia, and Algeria, added to those street protests in Spain, Austria, France, Germany, England, and New Zealand. The demonstrations were so protracted—Hong Kong's started in June 2019 and has gone on week-ly since—and so widespread that in the middle of it all the *New York Times* said that "2019 might be the year of the protest."

In fact, it is the opinion of two recent political scientists that "the state system seems to be failing all over the world," and they have proposed a new study called "archy" to examine how to grow, maintain, and fund states so as to avert their collapse. No better evidence of the seriousness of the world's "uproar and disenchantment" can there be when academics need to create a discipline to overcome it.

Another source of political turbulence, one that exacerbates all the rest, is overpopulation. Now with more than 7.7 billion people, the earth is in a condition ecologists call "overshoot," and as a matter of fact has been for some time: it is the state in which the drawdown of resources by one particular bipedal species, with technology honed to increase that drawdown as fast and efficiently as possible, has used up more than the earth can replenish, or ever will be able to replenish in thousands of years. The steady increase in consumers and the concerted effort to extend their lifetimes assures that the drawdown can continue indefinitely, while the amount of resources dwindles steadily and in some cases dries up altogether. Rare-earth metals, for example, which are by definition rare, have a clear end-time in the next 50 years maximum, and some, like iridium and platinum, may have only another 10 years; fresh water, too, is diminishing rapidly and it is calculated that 2 billion people will face water shortages in the next five years. We have already seen that animal resources have decreased by 80 per cent or so, and that is a rate of extinction that is not diminishing.

All that, of course, leads to competition and, in some cases, war, and since there will be no diminution in overpopulation—it has grown steadily and irredeemably by 83 million people *a year* since 1975—those conflicts can only increase. With the certain byproduct of displacement and immigration and thus the additional border clashes and violence. Add to that the failure of agriculture in an increasing number of locations as earth's temperature increases and water sources dry up, outdoor labor becomes impossible, and soils are exhausted.

One final evidence of political collapse is the failure of the single organization that attempts to represent the entire world, the United Nations.

It is too much to say that the U.N. is laughable because the immense waste of time, money, and energy that it has racked up in 75 years has been a global tragedy. The single purpose for which it was initially established—assuring peace in the world—it has not come anywhere near achieving as we have just seen, and its feeble attempts at it have only demonstrated how empty and empty-headed was the idea that it could. As to the ubiquity of war, everywhere on the globe, we have already seen that it has managed to reach a level not matched since the end of the war that begot the U.N. and at no time in its 75 years of existence has it been successful in heading off a threatened conflict. Its various peace-keeping attempts, sending blue-helmeted troops to wars already in progress—for example Darfur in 2003, Sri Lanka in 2009, to pick two of the most recent—has produced nothing but disasters. And it has stood helplessly by, even when charged with intervention, while there have been well-known cases of genocide, including Bangladesh in 1971, Rwanda in 1993-4, Serbia in 1995, Burma in 2018.

And in the one area outside of peacekeeping that it has taken upon itself to regulate and solve—global warming—it has managed to spend an immense amount of time and money, issuing lengthy scientific reports on the dire future, putting on expert forums with people from around the world, establishing an international convention on climate action, and achieving nothing: the great and mighty Oz but don't look behind the curtain. It has after 15 years of scientific inquiry finally come out in 2018 with a report that says in effect that everything will be catastrophic in 12 years the way we're going, and then put on a worldwide conference in 2018 at which nothing was done in any serious way to stop that tragedy, followed by another in 2019 with similar pitiable effect.

The cost to the world of this failure—to the United States,

mostly—has come very close to a trillion dollars over these years. Its official budget has exceeded half a trillion—the 2018-19 figure was $5.4 billion, for example—but there is in addition a peace-keeping budget that has run to about $400 billion, including $6.7 billion in 2018-19. This is sufficient to maintain a huge bureaucracy, and at very high levels, that somehow finds makework to do without worrying about the utter uselessness of its labor. But not sufficient to effect any essential changes in the way the world operates, certainly nothing approaching the charges of its charter.

The United Nations, in short, is an example of the collapse of politics at the global level. It is not too much to say that it is an outstanding example of the failure of the project created after World War II with the aim of spreading a Bretton Woods free-trade capitalism around the world based on Western-style liberal "democracies," aimed at a global market and even a global government, remaking the entire world in an American image. Not only has that not happened, but capitalism has caved in on itself and the liberal dream has everywhere turned into a disputative autocracy or a failed anarchy.

Another global institution that is unravelling in the 21st century is the Catholic Church. Not yet divided and discredited everywhere—it has after all lasted two millennia because it has been adroit at survival—it has proven itself incapable of either self-reform (in the case of the widespread sexual abuse scandals) or doctrinal coherence (in the contest between the marxicized Franciscans and the orthodox Benedictines). As of 2020 it stands as a divided and incoherent church in a similar world, rather than as the beacon of reason and comfort such a world needs. Pope Francis was right to say that "doomsday predictions can no longer be met with irony or disdain," and his church is powerless to provide answers…or solace.

A healthy political fabric of a society depends of course ultimately upon a healthy social fabric, and it is no surprise to find that the world's social ecology has deteriorated along with its physical ecology.

The breakdown of family and community, not only in the industrial world but increasingly in regions of poverty, hunger, and failed states, has been decried for generations now, without any system or faith or savior coming along to restore what is irretrievably buried underneath the asphalt of global capitalism. That means that the individual is all too often left on its own to fend amid the onslaught of a media-saturated and internet-permeated world, battered by economies that everywhere are increasingly polarized between rich and poor, and it is no surprise to find that the rates of suicide, addiction, and mental illness have been growing, in some places alarmingly so, quite around the globe in the last decade. To take just the United States, suicides and drug overdoses have gone up so sharply since 2015, now reaching some 3 million deaths a year—more than at any time in U.S. history—that they have driven down the population's life expectancy *five years in a row,* an extraordinary and unprecedented figure for a developed nation not at war.

To continue with the U.S., one way to measure the shredding of the social fabric is to remember the kinds of values that sustained this country for the longest time, from the founding until at least 1960: the stability of marriage, the embrace of religion, the importance of work, and participation in civic life.

*As of 2020, the institution of marriage is in a shambles, with only half the adults 18-64 married (and half of those are without any partner), the lowest figure ever recorded, 36 per cent have never been married, and a third of all births are outside marriage. (In the rest of the world, marriages declined from 7-10 per 1,000 in the 1970's to 5-7 in 1990's to 3-4 in the 2010's.) The divorce rate in the U.S. has climbed steadily from 1950 for all age groups, though recently younger people tend to have fewer divorces because there are fewer marriages.

* Religious affiliation and belief has declined in America since the middle of the last century, and a *Washington Post* report in 2018 cited studies showing that "religion is fading away" for most Americans,

especially younger adults, and that "markedly fewer Americans participated in religious activities or embraced religious beliefs." A Pew study in 2018 found that "Nones"—unaffiliated, agnostics, atheists—were more numerous than any single affiliation but evangelical Protestants, outnumbering Catholics, Muslims, Jews, or any fringe church. And another Pew study in 2019 said simply that the "decline of Christianity continues at a rapid rate." America may still be a Christian country no matter how the left may decry it, but less so—and less ardently, seriously so—with every passing year.

As religion fades, so of course does morality. American morality in the last two decades has been in such decay that certain elements are so common in the worlds of entertainment and celebrity that they no longer shock. The years-long practice of pedophilia in the Catholic church, accepted and hushed-up by the hierarchy, has been called "the worst crime in American history," given its scale, duration, and severity at all levels of society. Similar sexual abuse by doctors, coaches, and other people in authority on a similar scale has been uncovered in recent years at institutions that had seemed above reproach. The #Me-too movement revealed widespread sexual abuse, not excluding rape and physical injury, also in prominent male figures, and the Jeffrey Epstein scandal showed that illegal manipulation of underage girls was common enough to be accepted in leading political, financial, and social circles. And all this taking place amid a swirling morass of confusion and incoherence about sex and gender and identity, in which the marital union of people of the same sex is regarded as normal and commendable and webgiant Facebook lists no fewer than 56 genders that it recognizes.

* The status of work has changed drastically in the last half-century. The standard middle-class jobs have diminished steadily, going along with the hollowing-out of the middle range of society, or what the *Financial Times* in 2016 described as the "steadily eroding" middle class. The ugly truth at bottom is that the global economy foisted on us by the liberal elite really doesn't need many people to function and

can be served in most cases by machines, now grown so sophisticated as to handle almost all manual tasks—and with artificial intelligence rapidly expanding, estimated to put half of all American jobs at risk in the next decade.

So it is natural that almost all surveys of job satisfaction show that more than two-thirds of those employed in the U.S. hate their jobs: a 2017 report from the Mental Health America outfit found that 71 per cent are actively looking for a job or contemplating it; a Gallup poll in April 2017 found two-thirds were "disengaged" from their job and do a minimum amount of work to get by; another survey that year found that more than 70 per cent of workers were dissatisfied with their "career choices." Most of the reasons had to do with the dull and repetitive nature of the jobs, especially at the middle and lower income levels, and with worries about lack of job security and inadequate health benefits. It didn't help that the actual effective income declined over the decades: the median household income in the U.S. in 1990 was $53,000, in 2018 $62,000, during which time the inflation rate rose by 3 per cent a year, or 87.5 per cent, and the equivalent of a 1990 salary would be $99,398 in 2017, $37,000 short. Not much to like.

* What Robert Putnam called "social capital" in his 2000 study *Bowling Alone* has declined even further than the grim picture he painted back then. His argument in the 1990's was that television and the internet (and its video games) had the effect of "individualizing" people's lives so that they had far less social contact than before—hence the decline in all civic organizations from the PTA to the Elks, the decline in volunteers of all kinds, and the replacement of bowling leagues by individuals bowling by themselves. That was before smart phones came along in 2007, and "social media" (Facebook in 2004, Twitter in 2006), and a development of self-absorption and narcissism so pervasive that it quickly became a public worry for educators, politicians, parents, and psychologists, among others.

Less talked about but similarly pervasive has been a steep decline in attitudes toward the civic institutions of government. A general distrust of, and contempt for, government has led to a drop in election turnout at all levels, party gerrymandering has turned most state governments into unchangeable swamps unresponsive to public will, and at the national level the U.S. Congress has never earned more than 20 per cent approval ratings for the last 20 years. The standoff over the wall on the Southern border and government shutdown in 2018-19 brought opprobrium with government to its highest level in decades. Not only do most people not participate in civic affairs, most do not think much of those who do.

The pinnacle of public distrust with government occurred in the turbulent years of 2017-2019, when it became obvious that a portion of the political establishment, including the intelligence apparatus—what was quickly termed the "deep state"—was acting on its own, secretly and illegally, to disrupt first Donald Trump's campaign and then his presidency. The operation, traced to the Obama White House, CIA, and FBI, ultimately failed, but not without discrediting the Justice Department and much of the Democratic Congress and creating a profound sense of disillusionment that will likely lead to a Trump reelection, should the union survive.

It is not surprising that political collapse spreads in many forms in a nation whose social fabric is distending so deplorably, a phenomenon now found throughout the developed world, matching the failure of states in the developing world. The so-called populist, or nationalist, movements that have emerged throughout the developed world—evidenced by the Trump victory and sustained activism in the U.S., the Brexit vote in Britain that the stodgy establishment tried to scuttle, the *gilets jaunes* in France and their counterparts in Belgium, Poland, and elsewhere in Europe—these movements represent the fundamental inchoate dissent from the liberal globalism that has been imposed since World War II. It is not civil war and armed uprisings and revolution everywhere, but those are now no longer unimaginable.

If I say they are inevitable I am in a minority—but a growing one. Hence it turns out that in 2016 the U.S. military, in the form of a training video created by the Army for U.S. Special Operations Command, is already preparing its armed forces for dealing with the collapse of "megacities" and moving in to prevent "urban terrorism"— in other words martial law. A government that has for years on its own soil engaged in a war on drugs, a war on illegal immigration, a war on gun violence, and a war on terrorism is well-prepared to have a war on disruptive populations in its backyard.

Jem Bendell: "The starting point for a generative discussion of the deep adaptation agenda is a difficult one. Because to begin to rigorously and imaginatively discuss this topic first requires us to accept the likelihood of near term societal collapse. By which I mean that within ten years, in whatever society we are living in, that we will find ourselves in a situation where our normal means of income, sustenance, security, pleasure, identity and purpose all disappear. As it is impossible to predict the future within complex systems, 'ten years' is not my prediction, and I mention it as a device to help focus this discussion without making people run out the room to stock their bunker. Please note that I am not suggesting we have the whole ten years: we might have less than that."

IV.

———～～———

ECONOMIC COLLAPSE:
A PERFECT STORM

Now that it's 2019, we're going to start the new year here at Peak Prosperity by responding to the wishes of our premium subscribers and making our most recent premium report free to everyone.

For those unfamiliar with our work, it's based on the idea that humanity is hurtling towards a disaster of our own making. Several powerful and unsustainable trends are all converging towards an ever-narrowing gap in the future.

Because of this, the individual and collective choices we make today take on ever-increasing importance. Our collective choices -- around such issues as rampant money-printing by central banks, the failure to wean ourselves off of fossil fuels, and tossing an entire younger generation under the bus because that's most convenient for an older generation afraid of living within its actual means

-- are all pointing to a diminished and disappointing future. We need to make better choices that align ourselves with these (and many other) looming realities.

--PeakProsperity.com, 2019

Any economy inextricably dependent upon relentless growth, and a concomitant commitment to consumption, is doomed to fail eventually in a finite earth with finite resources. And in the process it is bound to create environmental disasters, political turmoil, and in the end economic chaos. Capitalism is that economy in its most successful and literally tragic manifestation, and its collapse is inevitable as, like the Sophoclean and Shakespearean heroes who go into disasters unable to change, we are unable to alter what we do.

The collapse has already started and many people beyond those at PeakProsperity tell us so, but it has not reached its ultimate disastrous end. We can see by the calamitous state we're in, however, at the beginning of the year 2020 what the reasons are. Let me spell out the principal ones:

1. Debt.

According to the Institute of International Finance, the world's debt stands at an unprecedented $247 trillion (a trillion is a million millions, lest we forget), a figure 318% bigger than the world's gross domestic product. Which is an economist's way of saying that the entire globe is living a pipe dream, and the awakening won't be long in coming.

The case of the United States, the world's largest economy for what that's worth as well as the world's most indebted, is indicative of the crisis. The official government debt was at nearly $23 trillion at the start of 2020, after more than 60 years of increasing obligations and a three-fold increase since the 2008 recession. That's an amount no less than three times as great as all the goods and services the economy creates each year—in other words, we can't make enough money to pay off the debt we already have. And no one doubts that the debt will only grow, year by year—nothing but blather, after all, is done to reduce it.

But wait—that's only the nominal debt, the existing Treasury debt. It does not include what according to generally accepted accounting principles are the unfunded *liabilities* of the government, the long-term obligations tied to Social Security and Medicare that amount to more than $100 trillion, or five times more than the publicized debt. When central banks start to worry about those numbers, the dollar is not likely to stay a stable currency.

Then add to that household debt, which stood at almost $14 trillion in 2019, nonfinancial corporate debt of $9 trillion, student loan debt of $1.5 trillion in 2019, obligations by the individual states (mostly in pensions) of $1.2 trillion, and debts by 63 of the largest cities of $330 billion—all of which stood at record levels at the end of 2019. The picture is perfectly clear: the entire economy is fueled not by profit, not by production, but by debt. Debt, moreover, that no one is working on paying back and that everyone needs to believe—this is the pipe dream that sustains the whole charade—will certainly be paid by future generations.

2. Dollar vs. Gold.

Despite the confident blather by economists and politicians at the time, the decision by the U.S. government in 1971 to go off the gold standard and let the dollar be the world's reserve currency, backed only by the good faith of the American government and a belief in an unfailing American economy, was bound to be unsustainable eventually. That would seem to be happening now.

Most of the central banks in the world have been getting rid of the dollar and buying gold, and the volume of gold buying in 2018 and 2019 was the greatest since the 1971 end of the gold standard, 651.5

tons in 2018 alone. China has been increasing gold holdings every year since 2017 and now holds nearly 2,000 metric tons (just over an American ton); Russia has added to its gold holdings at an increasing pace, adding 78 tons in the first months of 2019 and now has 2,135 tons; Germany, Italy, and France added to their reserves as well. It was, as Forbes magazine said, "a gold-buying spree" from the beginning of 2018 on.

And of course all of this gold was bought by U.S. dollars (Russia liquidated 90 per cent of its Treasury bonds in 2018, increasing its gold reserves from $14.9 billion to $61.9 billion), and thus the dollar began to lose its primary place as the world's reserve currency. Britain, France, and Germany set up the Instex system in 2019 as a way to trade outside the dollar system, primarily, and China and Turkey did the same. As a result, the dollar continued its steady decline in value, losing almost 98 per cent against gold from 1971 to 2019, and the decrease shows no sign of halting. One Swiss banker in 2018 in fact said bluntly, "The long-term trend of the dollar is clear: it will go into oblivion faster than anyone can imagine."

So the world's primary economic question now has become: who has how much in gold? There is no easy answer to that, because each nation reports its own holdings and there is no overarching authority to guarantee their accuracy. The World Gold Council reports figures annually that show Germany, Italy, and France to be behind the U.S., followed by Russia, China, and Switzerland, but they only use the figures the individual nations supply them. The conventional wisdom is that the U.S. has the largest reserves, at 8133.5 metric tons in 2019, which at a rate of $1200 an ounce would amount to $320 billion. The trouble is, this is the same amount that the U.S. declared in 1971 and no one thinks that the country has not spent any of its gold in 50 years; in fact a Swiss banker asked in early 2019, "Is there any U.S. gold left?" without receiving an answer.

I suspect we may find out this year.

3. *Limits to Growth*

In 1972 an outfit called the Club of Rome published an analysis called *Limits to Growth* that showed by computer analysis what the kind of growth that we have seen in the last 30 years—exponential, heedless production—would do to society. It concluded that "if the present growth trends in world population, industrialization, pollution, food production, and resource depletion continue unchanged, the limits to growth on this planet will be reached sometime in the next one hundred years. The most probable result will be a rather sudden and uncontrollable decline in both population and industrial capacity"—in fact, not to put too fine a point on it, "overshoot and collapse" in all dimensions with a tipping point around 2020-30.

It was largely ignored. Growth is, after all, the lifeblood and raison d'etre of capitalism and the idea that there should be limits to it was so repugnant as to be essentially incomprehensible. A lot of people are making a lot of money, and a very few a very lot, so what would be the point in imagining that there would ever be an end to it. And so capitalism became global capitalism, and then added computer capitalism and internet capitalism, and it continued to grow exponentially with no thoughts whatsoever to limits.

A second study in 2000 found that the original analysis thirty years earlier could not be invalidated. "To the contrary," said international banker Matthew Simmons, "the chilling warnings of how powerful exponential growth can be are right on track." Capitalism still ignored it and added cellphone capitalism and central bank appeasement into the mix and went right on growing.

The third study came in 2014 at the end of the Great Recession from University of Melbourne scholar Graham Turner, who came to the conclusion that resource depletion, overpopulation, economic decline, industrial slowdown, and environmental collapse were following almost exactly the lines of the 1972 forecast—and leading straight to "overshoot and collapse" in 2020-30. A group called the Global

Footprint Network was started in 2003 to measure that overshoot annually. It figured that the world was using up 1.7 Earths in 2015, closer to 2 Earths in 2019—in other words, we are growing at a rate so rapid that we are using up the resources of the earth twice over— and polluting the atmosphere twice as much—every single year. No wonder collapse.

4. Financial Inequality.

On a global level, it has been calculated that the twenty-six richest men in the world have a net worth equal to the poorest *half* of the world, 3.6 billion people an inequality that Oxfam has said "has reached extreme levels." Nothing much more to say after that except to note that global warming will affect the poorest half far more severely than richest two dozen.

As to the U.S., it ranks in the bottom 30 per cent of nations in terms of inequality of incomes—in other words, 70 per cent of the world has economies that are more equal than America's. But since America has the world's biggest economy, that makes a difference, and in 2019 its inequality was at the highest level it has ever been: the top 1 per cent earns an average of more than $6.5 million a year, 188 times as much as the bottom 90 per cent, the top 10 per cent make 9 times as much as that bottom 90 per cent.

The problem is easily stated. The average salary for American executive officers as of May 2019 was $14.5 million, the fourth straight year the salaries set a record, 287 times more than their workers' average. Median household income at the same time was $63,000, also a record, but it was the same in purchasing power as the top household income in 1999, inflation having taken its regular bite. Another way of seeing it is looking at average hourly earnings, which in 2018 had less purchasing power than in 1973, for the same reason. Or again: hourly earnings increased from 1990 to 2018 by nearly 12 per cent—and drugs increased by 182 per cent, medical care by 184 per cent, food by

278 per cent, and college tuition by 374 per cent.

In the history of the world, it has not always been the contrast between rich and poor that has led to uprisings and revolutions, because often the downtrodden are indeed trodden down so far as to be incapable of resistance; it is often a rising economic sector that wants to take over power from the settled rich. But taken together with the other elements of economic fragility such as those outlined above, such inequality seems certain to provoke the kind of political unrest that Plutarch had in mind 2,000 years ago when he said that "an imbalance between rich and poor is the oldest and most fatal ailment of all republics." All the elements of collapse are there before us, and all may be said to be at work. It cannot be long before the bough breaks.

There was no room at all, in these ways of thinking, for the novel, apocalyptic situation which had now arisen, a situation which needed solutions as radical as itself. (The Status Quo) attitude is a complacent acceptance of things as they are, without a single new idea.

This acceptance was accompanied by greatly excessive optimism about the present and future. Even when the end was only sixty years away, and the Empire was already crumbling fast, Rutilius continued to address the spirit of Rome with the same supreme assurance.—Michael Grant, *The Fall of the Roman Empire,* Barnes & Noble, 2005

V.

———

CONCLUSION...?

SO IT LOOKS as if I have lost the bet.

Not by much, obviously, and only insofar as my arbitrarily selected date was off by a few years. I would say that the consensus today is that it was off by perhaps a decade, not much more than that. The unchecked rampages of technology have gone on, and intensified, just as I predicted they would, with consequences of much greater climate crises, much more turbulent political systems, much more fragile global economies. Heedless technological advances pushing heedless exponential growth beyond human capacity to control stands guilty, just as I predicted. And technological miracles to solve or reverse this onslaught—carbon-capturing, solar modification, and the like—never emerged, because that is not the direction in which the overriding logic of modern high technology moves. (Not to mention technological miracles by which scientists infected with MITitis believe we will be saved by uploading ourselves onto artificial-intelligence machines.)

I acknowledge that to say that the collapse of industrial civilization will happen ten years later than predicted is still to validate the prediction beyond anything that Kevin Kelley would have believed. Still, he is welcome to cash the check I put in the hands of our mutual editor those 25 years ago if he should wish to.

The reason that the collapse of human civilization does not trouble me overmuch is that I am an ardent believer in Gaea, the mother earth, creator of the creators, first of the cosmos, in the words of Plato, "a living creature, one and visible, containing within herself all living creatures." And I believe that earth will continue to exist for billions of years until the sun explodes and the Solar System is no more.

That there will likely be many fewer human animals, and none with a desire to build more civilizations, can be viewed only as a blessing, as Gaea would see it. Those civilizations, after all, were the ones that led up to that great destructive process of *human domination* that brought about the collapse, and the disruption of the plant and animal worlds and extinction of species she had taken so long to create since the last cataclysmic destruction of life 66 million years ago.

Mother Nature, as the saying has it, bats last.

I am of course deeply troubled that those I know and love may not continue to exist, that many benevolent and righteous souls may vanish, and that much that was good in human achievement and art is gone forever. That is nothing but sad, yea heartbreaking, for any sentient human. But when I reflect upon what that species has done, what fundamental evil it has thrust upon the world to cause its civilization to collapse, I cannot but feel that it is ultimately for the betterment of Gaea that it should cease to thrust its pernicious presence upon the earth.

Twenty-five years ago I felt that, I feel it now. I only hope that if there are small populations of humans that exist in future years, and there are enough remaining species of plants and animals to sustain them, that they will have learned the lessons of civilizations out of control, in thrall to ideas of growth, and technological improvement, and human superiority, and never go down those paths again. After all, humans lived for most of two million years without any thought of perpetual improvement and development, without any technologies that would see to those ends—the simple hand axe sufficed human-kind for most of that time—and without the mad idea that the earth, and all the creatures thereof, was made primarily for human use and enjoyment.

But I have learned over a long life no longer to hope. Hope, after all, was the gift from Pandora that did not escape her jar.

NOTES

All of the citations are easily available on the internet, chiefly at Wikipedia (cited here as w.p.).

I. THE INCIPIENT BET

Rebels Against the Future: The Luddites and Their War on the Industrial Revolution: Lessons for the Computer Age, Addison-Wesley, 1995, pp. 235-36.

Kevin Kelly, *Wired,* 6/1/1995.

Union of Concerned Scientists, Washington, 1992.

Jem Bendell, w.p.; Deep Adaptation: A Map for Navigating Climate Tragedy, iflas.info or lifeworth.com/deepadaptation.pdf.; carbon emissions, *National Geographic,* 9/18/19.

II. ENVIRONMENTAL COLLAPSE

Ecosystem Millennium Assessment, *Science,* 2005.

Pope Francis, papal encyclical Word on Fire, https://laudatosi.com/watch.

2017 Union of Concerned Scientists, ucsusa.org/sites/default/files/attach/2017/11.

Ten hottest years: NOAA.gov/news, 2/6/19; USA Today, 7/18/19.

Melting ice: NOAA Arctic report, 12/11/18; AP 12/12/18; UN Intergovernmental Panel on Climate Change (IPCC), *Science,* 1/10/19; World Glacier Monitoring Service (wgms.org), Global Glacier Change Bulletin (wgms.ch/ggcb); Peter Wadhams, *A Farewell to Ice* (Alan Lane, 2017), and realscience.com/2015/10; *The End of Ice,* Dahr Jamail (2019).

Albedo effect: *The Verge,* Alessandra Patenza, 5/10/18.

Particulate pollution: J. Marvin Herndon, *Journal of Geography, Environment, and Earth Science,* 17(2) 1-8, 2018, Ian Baldwin, *Vermont Commons,* 2018-19, and globalresearch.co/ianbaldwin&x=14&y=C10.

Sea levels: IPPC, *Science,* 1/10/19, Angela Fritz, *Washington Post,* 1/13/19, Laure Resplandy et al, *Nature,* 10/31/18; Michael Oppenheimer, AP 9/26/19, IPPC report to UN, 9/25/19.

Extreme weather: Center for Climate and Energy Solutions (c2es.org/content/extreme-weather-and-climate-change) and (national-climate-assessment), 2019.

Methane: Arctic-news.blogspot.com; Arctic Methane Emissions, w.p.; Arctic Methane Emergency Group, Facebook, and Shakhova interview 6/24/17; Shakhova et al, *Geosciences,* 6/5/19; Chris Mooney, *Washington Post,* 3/19/18; Wilkerson et al, *Atmospheric Chemistry and Physics,* June 2019; archaeologynewsnetwrk.blogspot.com/2018/09/greenhouse-emmissions-from-siberian.html#cxKF8A21BXgorYbJ.

Oceans warming: Chris Mooney, Brady Dennis, *Washington Post,* 11/l/18, on Laure Resplandy, op. cit.; Kevin Renberth et al, *Science,* 1/10/19; IPCC report to UN, 9/26/19.

Oceans acidity: James Bradley, themonthly.com>author>james-bradley, August 2018, and *National Geographic,* August 2018; Frances E. Hopkins et al., PNAS, pnas.org/conent/107/2/760, 1/12/10; science2017.globalchange.gov/chapter/13; Bendell, Deep Adaptation, op. cit. Vernon, cnn.com>2018/08>world>great-barrier-reef.

Extinctions: UN Intergovernmental Science-Policy Platform on Biodiversity and Ecosystem Services, May 2019, and Associated Press, 5/7/19; PNAS, Gerardo Ceballos et al., 7/10/17, CBS News 7/10/17; nature.com/articles/s41598-018-35068-1; Bradford Lister et al., PNAS 2018 doi.org/10.1073/pnas.1722477115; Elizabeth Kolbert, *The Sixth Extinction*, Picador, 2015.

NTHE: facebook.com.groups/NTHEsupportgroup and NTHEevidencegroup.

III. POLITICAL COLLAPSE

Wars: warsintheworld.com; List of Ongoing Armed Conflicts, w.p.; Council on Foreign Relations, Global Conflict Tracker; ourworldindata.org/war.

Syrian drought: Environmental Issues in Syria, w.p.; Agriculture in Syria, w.p.; vice.com>en_us>article>the-drought-that-preceded-the-Srian-civil-war.

IPCC Report, *N.Y Times,* August 8, 2019, p. 1.

Red Cross, October 30, 2018.

Failed states: worldpopulationreview.com>countries>failed-states; fundforpeace.org>2019/04/10>fragile-states-index-2019.

Kunstler: Lewrockwell.com, 6/18/19.

Protests: *N.Y. Times,* 6/25/19.

Overpopulation: Human population, w.p.; weforum.org/agenda/2018/10/david-attenborough-warns-planet-can'tcope-with-overpopulation.

United Nations, w.p., Criticism of United Nations, w.p.; un.org.; quora.com>is-united-nations-a-useless-orga

Pope Francis: papal encyclical Word on Fire, 6/15, https:laudatosi.com/watch.

Drug overdoses, suicides, life expectancy, Centers for Disease Control and Prevention report, 11/29/18, AP11/29/18.

Marriage and divorce: Marriage in the U.S., w.p., Divorce in the U.S., w.p.

Washington Post, 5/15/18; Jean Twenge, SGWPUB.COM,, 3/23/16. Pew report 2018, Pew Research Center, 8/8/18 and 10/17/19, pewresearch.org.

Work: *Work in America: Report of a Special Task Force, 2/15/*; *Financial Times,* quoted in *USNews,* 5/27/16; Mental Health America, *Mind the Workplace,* 2017; CBS News, 4/3/27; Gallup Poll, in *Forbes,* 9/22/17.

Robert D. Putnam, *Bowling Alone,* Simon & Schuster, 2000, revised and updated, 2020.

Bendell, op. cit. (Ch. I).

Training Video, report for U.S. Army War College, theintercept. com>2016/10/13>pentagonvideo

IV: The Economic Collapse: A Perfect Storm

Debt: Institute of International Finance, Global Debt Monitor, 3/19; Truthinaccounting.org (Our Debt Clock, Financial State of Union, Cities). Other debts: *Fortune,* 1/1/19, NY Federal Reserve,12/19.

Dollar vs. Gold: The Great Debasement, Zero Hedge, 2012, zero-hedge.com/news/2019-09-04/great-debasement; World Gold Council (www.gold.com); RT News, Lewrockwell.com., 6/2519; Gold reserve, w.p. Swiss banker: Egon von Greyerz, Lewroockwell. com., 2/7/19.

Limits to Growth: Donella H. Meadows, et al., *The Limits to Growth,* Signet Books, 1972, 1975, Chelsea Green, 2004; Matthew R. Simmons, 10/2000, greatchange.org/ov-simmons, club-of-rome-revisited. html; Graham Turner and Cathy Alexander, *Guardian,* 9/1/14, https://theguardian.com/commentisfree/2014/sep/02/limits-to-growth-was-right-new-research-shows-were-nearing-collapse; Global Footprint Network (foootprintnetwork.org).

Financial Inequality: Twenty-six richest men: Oxfam report, 1/16/17,

1/20/19, vox.com, 1/28/19, global warming, Noah Diffenbaugh, et al., Proceedings of the National Academy of Sciences, 3/22/19. U.S. inequality, AP, 9/27/19. Executive pay, vox.com/policy-and-politics/2019/6/26/18744304/ceo-pay-disclosed. Purchasing power, Fortune 1/1/19.

INDEX

About the Author

Kirkpatrick Sale is an independent scholar and the author of fourteen books, including *Human Scale* (Coward McCann) and *Human Scale Revisited* (Chelsea Green), and *SDS* (Random House), *Conquest of Paradise: Christopher Columbus and the Columbian Legacy* (Knopf), *Rebels Against the Future: The Luddites and Their War on the Industrial Revolution* (Addison Wesley), *After Eden: The Evolution of Human Domination* (Duke University), and *Why the Sea Is Salt: Poems of Love and Loss* (iuniverse). He has been a contributor to magazines and newspapers from coast to coast, including *the New York Times, Los Angeles Times, Charleston Post and Courier, Newsweek, The Nation, Mother Jones, New York Review of Books, The American Conservative, Chronicles, Vermont Commons,* Counterpunch.com, and Lewrockwell.com. He has written introductions to Leopold Kohr's *Breakdown of Nations*, E. E. Schumacher's *Small Is Beautiful,* Tobias Lanz's *Beyond Capitalism and Socialism*, and Thomas Naylor's *Secession* and contributed essays to more than twenty other books. He was an officer of the PEN American Center for many years, was a founder and officer of the E.F. Schumacher Society, and is currently director of the Middlebury Institute for the study of separatism, secession, and self-determination. (MiddleburyInstitute.org). He is in *Who's Who in America* and a profile appears in Wikipedia (Wikipedia.org/wiki/Kirkpatricksale). He lives in Mt. Pleasant, South Carolina.